I0493412

SUCCEED MY TEST MATHEMATICS GRADE 7
By R CHENJE

First Printing: 2014

ISBN: 978-1534949798

Chentronics Publishers
4945 Mkoba 11 Gweru Zimbabwe

+263 771 303 686

U.S. trade bookstores and wholesalers: Please Contact Chentronics Publishers, 4945 Mkoba 11 Gweru Zimbabwe

Mobile: +263 771 303 686

Or email pstchenje@gmail.com

Contents

1.

H	T	U	t	h
• • •	• • • • •	• • • • • • • • •	• • • •	• •
• • • • • •	• • • • •	•	• • • • •	• • • • • • •

The abacus shows

a. 640,57

b. 35,842

c. 64,057

d. 358,42

2. Write the value of the underlined digit 69<u>3</u>4

   a. 9   b. 90   c. 900   d. 9000

3.   4c = $

   a. $4,00   b. $0,40 c. $40,00   d. $0,04

4. $\frac{594}{6}$ =

   a. 89   b. 99   c. 79   d. 59

5. 1 - 0,550 =

   a. 0,450   b. 0,549 c. 0,550 d. 0,540

6. $2\frac{3}{8}$ as an improper fraction =

   a. $\frac{16}{8}$ b. $\frac{13}{8}$ c. $\frac{19}{3}$ d. $\frac{19}{8}$

7. 0,05 + 3 =

   a. 0,08   b. 0,15 c. 3.05 d. 3,50

8. Find the HCF of 12; 15 and 27

   a. 12   b. 27   c. 3   d. 54

9. Which of these numbers is a prime number?

   a. 2   b. 6   c. 9   d. 15

10.   4368g = _____ kg

   a. 43,68   b. 4,368   c. 436,8 d. 4368

11.

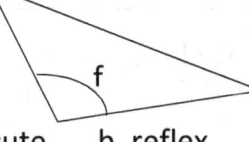 angle f is _____

   a. acute   b. reflex   c. obtuse   d. right- angled

12.   The value of 3^3 + 3is

   a. 9   b. 27   c. 12   d. 30

13.   Find the factor of 42

   a. 7   b. 8   c. 9   d. 5

14. Put >, <, = or ≠ $\frac{4}{5}$ ☐ $\frac{9}{10}$

   a.>   b.<   c.=   d.≠

15. $\frac{4}{9}$ is the same as

   a. $\frac{8}{27}$ b. $\frac{8}{18}$ c. $\frac{16}{54}$ d. $\frac{16}{27}$

16.   Complete the sequence 6; 9; 14; 21;☐

   a. 24   b. 30   c. 28   d. 27

17.

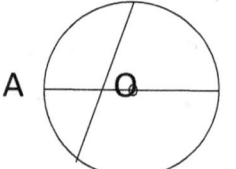

A   Q   B

line OB on the circle is the _____

   a. radius   b. circumference   c. chord   d. diameter

18.   906x38 =

   a. 39330   b. 34428   c. 34430   d. 39334

19.   Morning

                The time shown on the clock is

                a. 1.05p.m

                b. 5.o5p.m

                c. 5.05a.m

                d. 5.01am

20. Express 0,65 as vulgar fraction

   a. $\frac{13}{20}$ b. $\frac{3}{20}$ c. $\frac{13}{25}$ d. $\frac{4}{5}$

21. 4cm /\ 4cm   Find angle x

   a. 45^0 b. 44^0 c. 30^0 d. 60^0

   4cm

22.   Find the average of 10, 12, 5, 7,1

   a. 5   b. 7   c. 10   d. 12

23. $10^3=$

   a. 300   b. 100   c. 1000   d. 103

24.   Which statement is false ?

   a. $0,2=\dfrac{1}{5}$ b. 1 ½>0,75   c. 0,5>¾   d. ¾<1

25. $\dfrac{2}{3}+\dfrac{3}{5}=$

   a. $\dfrac{14}{15}$ b. $\dfrac{5}{8}$ c. $\dfrac{9}{5}$ d. $\dfrac{5}{15}$

26.   6 March 2011 was on Wednesday. What day was

   27 March 2011?

   a. Monday b. Sunday   c. Wednesday   d. Sunday

27.   The value of 7 in 1,957km is

   a. 7m   b. 7km c. 7cm d. 7mm

28.   _____ is a square number.

   a. 2   b. 3   c. 4   d. 5

29. Express 36 out of 72 as a percentage

   a. 50%   b. 36% c. 72%   d. 5%

30. $\frac{8}{9} \div 4 =$

   a. $\frac{2}{3}$ b. $\frac{2}{9}$ c. $3\frac{5}{9}$ d. $3\frac{2}{9}$

31. $2,60 to the nearest dollar is

   a. $2,00   b. $2,60   c.$3,00 d. $1,60

32. Calculate time taken to cover 400km at 80km/hr

   a. 4 hours   b. 8 hours   c. 5 hours   d. 2 hours

33. Find the larger amount when $20 shared in the ratio

   1:4

   a. $16   b. $5   c. $4   d. $1

34.   15 minutes afternoon is the same as

   a. 0015   b. 1215   c. 1500 d. 1512

35. Find the perimeter of an equilateral triangle with sides

   6,5cm.

   a. 6,5cm   b. 13cm   c. 19,5cm   d. 21cm

36.   0,345x100 =

   a. 34, 5   b. 3,45 c. 4500   d. 345

37. Tsitsi is travelling west. She turns through 3 right angles anti clockwise. She is now travelling _____

a. south   b. north   c. east   d. west

38. This shape has six faces

a. cylinder   b. triangle prism   c. sphere   d. rectangular prism

39. _____ is a multiple of 9

a. 33   b. 37   c. 74   d. 54

40. 2/3 of 45=

a. 30   b. 67,5   c. 90   d. 15

41. 23,539 to the nearest hundredth is

a. 23,50   b. 23,53   c. 23,54   d. 24,00

42. Find the volume of a box 11cm long, 8cm wide and 3cm high

a. 260cm   b. 33 cm^3 c. 264cm^3 d. 88cm^3

43. The buying price of a shirt was $69,37 and the selling price was $56,79. Calculate loss

a. $12,58   b. $81,95   c. $44,21   d. $35,84

44. Susan left ¾ of a cake in a tin. Rudo ate $\frac{1}{3}$ of it. What

fraction of the cake did Rudo eat?

a. ¾   b. $\frac{1}{3}$ c. $\frac{5}{12}$ d. ¼

45. Find the area of the triangle?

a. 30 cm^2 C. 60cm^2

b. 48cm^2 D. 80cm^2

46. A shape with six (6) sides is a _____

a. pentagon b. hexagon   c. quadrilateral   d. cube

47.   28 October 2011 in SI notation is

a. 2011/ 10/ 28   b. 2011/ 11 /10

c. 28/10/ 2011   d. 28/11/ 2011

48. Which two numbers have a sum of 17 and a product

of 72?

a. 15 and 12   b. 8 and 9 c. 10and 7   d. 13 and 4

49. Simplify $\frac{5}{8}$ x $1\frac{3}{5}$ ÷ $\frac{1}{8}$=

a. $\frac{1}{8}$ b. $\frac{13}{64}$ c. 8   d. 13

50. Eighteen minutes to six is _____

a. 6.42   b. 6.18   c. 6.12   d. 5.42

1.(a) $1\frac{5}{8} \div \frac{1}{4}$ [1]

(b) 33 979 to the nearest 1 000   [1]

2.

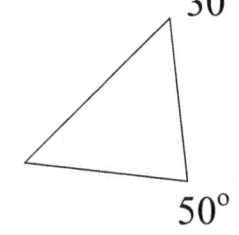

30°

X   50°

Find angle X   [2]

3.

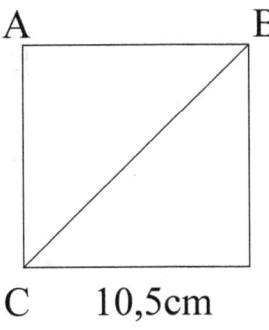

A   B

7,2cm

C   10,5cm   D

(a)   Write down the name of line CB   1]

(b)   Calculate the perimeter of the rectangle

4.Find the difference between 2 ¼ and 2 2/3

5.(a) (3 x 6) x 4   [2]

   (b) 11 − (2 + 3) =   [1]

6.Find the average of 76 and 88   [2]

7.(a) $9,05 x 3   [2]

   (b) Write 3 2/5 as an improper fraction   [2]

   (c) 1/3 + ¼   [2]

8.(a) What is the difference between 74 275 and

   19840?   [2]

   (b) What is 3/5 of 150 days?   [2]

   (c) 1 day has _____ hours   [1]

SECTION B

ANSWER ANY THREE QUESTIONS IN THIS SECTION

9.(a) ¼ + 6/12   [2]

  (b) Take away 6 from 18   [2]

  (c) 24 + 12   [1]

10.

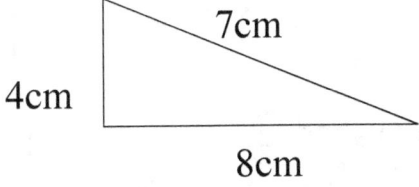

  (a)   Name this shape   [1]

  (b)   Find the area of this shape   [2]

  (c)   ½ past 4 in the morning in 12 hour

    notation   [2]

11.

  (a)   Name this shape   [1]

(b)   21,27 in expanded notation   [2]

(c)   2250 minus 40   [2]

12. (a) Add 4467 and 5783   [2]

  (b) 4 x (2 x 3)   [2]

  (c) 10 + 12   [1]

13.

(a)   What fraction is shaded?   [1]

(b)   49 002 in words   [2]

(c)   200 – 1   [2]

1. Seven thousand and thirty –nine is _____

    a. 739   b. 7 039   c. 7 390   d. 7 93

2.   563 to the nearest 10 is _____

    a. 570   b. 560   c. 565   d. 580

3. $\frac{13}{9}$ as a mixed number is _____

    a. $1\frac{3}{9}$ b. $1\frac{4}{9}$ c. $3\frac{1}{9}$ d. $9\frac{1}{3}$

4. $\frac{5}{10}$ as a decimal number is _____

    a. 0,5   b. 5,0   c. 1,2   d. 10,5

5. 834 + 99=

    a. 1824   b. 933   c. 9330   d. 1033

6.

2cm   Area of this shape is _____

    8cm

    a. 20cm^2 b. 10cm^2 c. 16cm^2 d. 4cm^2

7. In a bus there are 39men, 22 women and 15 children.

    How many people were in the bus altogether?

    a. 65   b. 41   c. 54   d. 76

8. A litre of petrol costs 97c. Mr Rusike puts 20 litres in his car. How much does he pay for the petrol?

a. $1940   b. $194,00   c. $19,40   c. $1,94

9. What is the product of 8 and 5?

a. 13   b. 3   c. 40   d. 24

10. Put the correct sign

72%   0,09

a.   >   b. <   c. =   d. U

11.   What fraction is shaded in its lowest?

a. $\frac{2}{12}$ b. $\frac{1}{6}$ c. $\frac{1}{3}$ d. $\frac{6}{12}$

12.   52,041 x 100=

a. 52,401   b. 520,41   c. 0,5241   d. 5240,1

13. Eighteen minutes to six is _____

a. 6.42   b. 6.18   c. 6.12   d. 5.42

14.   1 day is _____ hours

a. 24   b. 12   c. 48   d. 6hours

15. $\frac{1}{6}$ of 42

   a. 3   b. 12   c. 48   d. 18

16. $\frac{3}{8} + \frac{5}{8} =$

   a. 1   b. $\frac{8}{16}$ c. $\frac{2}{8}$ d. $\frac{16}{8}$

17.   1/6 + 2/3 =

   a. $\frac{6}{6}$ b. $\frac{5}{6}$ c. $1\frac{5}{6}$ d. $\frac{6}{5}$

18. 3cm 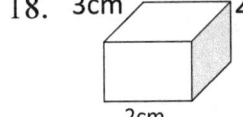 2cm   Find volume

      2cm

   a. 12cm^3 b. 8 cm^3 c. 18 cm^3 d. 6 cm^3

19.   A rectangular prism has _____ dimensions

   a. 12   b. 3   c. 8   d. 4

20.   1 litre is _____ ml

   a. 1000   b. 100   c. 500   d. 1500

21. A factor of 42 is _____

   a. 10   b. 4   c. 6   d. 8

22. Saturday is 29th November. What day is 12th December

   a. Saturday   b. Friday   c. Thursday   d.
   Wednesday

23. A car had 50 litres of petrol in its tank. On a journey 31 litres were used. How many litres of petrol were still in the tank?

   a. 81 litres b. 8 litres   c. 19 litres   d. 20 litres

24.   Line AB is _____

   a. circumference   b. diameter   c. radius d. cord

25. Which one is a right angle?

   a.   b.   c.   d.

26. One shirt cost $2, 00. What is cost of 6 shirts?

   a. $8   b. $4   c. $12   d. $1,20

27. The sun rises in the _____

   a. North   b. South   c. East   d. West

28.   3 $5 notes _____

   a. $8   b. $2   c. $15   d. $18

29.   6$\overline{)\$41,70}$

   a. $69,50   b. $6,95   c. $695   d. $9,65

30.   5c x 20 =

 a. 50c   b. 10c   c. $10   d. $1

31. $\frac{6}{8}$ = —

 a. ¾   b. ¼   c. ½   d. $\frac{1}{3}$

32.   5,77 to the nearest tenth is _____

 a. 5,75   b. 5,80   c. 6,80   d. 5,00

33.   1cm =   mm

 a. 100mm   b. 10mm   c. 15mm   d. 100cm

34. What is the value of '7' in 3 674?

 a. 0,7   b. 700   c. 7000   d. 70

35. The sum of $6,70; $5,40; and $3,30 is _____

 a. $1,54   b. $5,41   c. $15,40   d. $16,40

36.   1m – 0,43m=

 a. 0,57m   b. 5,7m   c. 0,057m   d. 570cm

37.   62
        x32
        —
                 a. 192   b. 1984 c. 94   d. 1824

38.   0.06 as a fraction is _____

 a. 6/10   b. 60/100   c. 6/100   d. 16/100

39. The value '9' in 36, 941 is _____

   a. 9 hundredths   b. 9 thousandth c. 9 units   d. 9 tenths

40.   1 week is _____ days

   a. 21   b. 7   c. 14   d. 28

41. $\frac{1}{5}$ of $1 is _____

   a. 20c   b. 2c   c. $5   d. 25c

42. $8 - \frac{2}{3} =$

   a. $7\frac{1}{3}$ b. $\frac{3}{22}$ c. $\frac{13}{7}$ d. $\frac{1}{3}$

43. U $\frac{1}{10}$ $\frac{1}{100}$ $\frac{1}{1000}$

                      The picture shows_____

   a. 3,59   b. 5,390   c. 3,059   d. 30,59

44.   875ml = _____ litres

   a. 37,5   b. 8,75   c. 0,875   d. 57,8

45. One litre of oil costs 72c. seven litres cost$_____

   a. $5,04   b. $4,50   c. $54,00   d. $45,00

46.   6,4 to the nearest whole number is _____

   a. 6   b. 10   c. 24   d. 7

47.
$$\begin{array}{r} 6000 \\ -\ 2875 \\ \hline \end{array}$$

   a. 2135   b. 3125   c. 1325   d. 2125

48. $2 \times \frac{3}{5} =$

   a. $\frac{16}{5}$ b. $2\frac{1}{5}$ c. $1\frac{1}{5}$ d. $\frac{10}{5}$

49. $\frac{7}{10}$ of 1000ml

   a. 70ml   b. 700ml   c. 7ml   d. 7000ml

50. $\frac{5}{6}$ of 1 day

   a. 12h   b. 14h   c. 20h   d. 16h

SECTION A
ANSWER ALL QUESTIONS

1. $3 + 4 + 2 = $ (1)

2. What is the difference between 20 and 15 (2)

3. Write 48 in words (2)

4. $39 = \square + \boxed{}$ (2)

5. $168 = \boxed{} + \square$ (3)

6. 2×9 equals   (2)

7.   2 2
     $-$ 5 (2)

8.   8   7
    $+$ 2   2   (1)

9. $4\,\overline{\smash{)}\,16}$ (1)

10. $19 \div 2$ (2)

11. $11 + 4 = $ (1)

12. Subtract 7 from 10   (2)

13. Find the sum of 3, 9 and 4 (2)

14. 20 divided by 5 (2)

SECTION B

13. James has 10 sweets. He eats 6 sweets. How many are left? (3)

[ ]

14   Name this shape.

How many corners does it have? (2)

15. 6 boys and 4 girls are in a family. How many children are they altogether (3)

16. Share 16 bananas equally between 2 boys. How many bananas does each get.

17. What is 25 to the nearest 10 (2)

18. How many cups in 10 boxes of 3? (2)

Instructions

1. Read all the instructions carefully

2. Answer all questions

1. Sixty thousand and eight in figures is
   a. 6008   b. 6080   c. 60008 d. 60080

2. The value of the underlined digit in the number 523, 1<u>8</u>6 is__
   a. 8 hundredths   b. tenths   c. 8 tens   d. 8 hundreds

3. 0, 75 can be written as
   a. $\frac{3}{40}$ b. ¾   c. $5\frac{7}{10}$ d. 7 ½

4. Round off 8,74 to the nearest whole number is _____
   a. 8   b. 9   c. 900   d. 870

5.   8cm

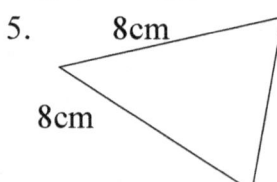

   8cm

The name of the shape above is

a. an isosceles triangle   b. an equilateral triangle

c. a scalene triangle   d. a right angled

6. 10:00 am in 24 hour notation is

    a. 10000   b. 1200   c. 1000   d. 2000

7. $\frac{6}{8}$ in its lowest terms is

    a. $\frac{12}{16}$ b. $\frac{2}{3}$ c. ¾   d. $\frac{4}{5}$

8. 0,31+4+0,84=

    a. 1,15   b. 1,19   c. 4,15   d. 5,15

9.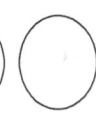

What is the number of cakes expressed as a mixed number?

    a. $3\frac{5}{6}$ b. $2\frac{5}{8}$ c. 3 ½   d. $3\frac{5}{8}$

10. What is 2005 cents in dollars?

    a. $200, 50   b. $20, 50   c. $20,05   d. $2,05

11. $2\frac{5}{12} + 3\frac{2}{3} =$

    a. $5\frac{7}{15}$ b. $6\frac{1}{12}$ c. $6\frac{5}{12}$ d. $6\frac{2}{3}$

12. Tendai was facing east. She turned on complete revolution and a half right angle in a clock wise direction. Then she is facing____

    a. south east   b. north east   c. north west   d. east

13. $ 4.50 as a fraction of $7,50 in its lowest terms is

   a. ¾   b. $\frac{3}{5}$ c. $1\frac{2}{3}$ d. $4\frac{5}{75}$

14. 2035 – 850+43=

   a. 1992   b. 1228   c. 1142 d. 2228

15. 64x 4 =

   a. 4^3 b. 2^6 c. 8^3 d. 4^4

16. Increasing 107 by 64 gives

   a. 747   b. 171   c. 63   d. 43

17. A shirt is marked $20. If 5% discount is offered for cash, what is the discount?

   a. $19   b. $15   c. $4   d. $1

18.

300cm

      3 000cm

The area of this rectangle in m² is

   a. 90   b. 66   c. 45   d. 33

19. $\frac{4}{6}$ of a day is =

   a. 36hours   b. 8 hours   c. 12 hours d. 16 hours

20. The perimeter of a square garden with sides 12cm is

   a. 144cm   b. 72cm   c. 48cm d. 24cm

21. Dividing the product of 8 and 40 by 16 gives

   a. 80   b. 64   c.42   d. 20

22. $16 \div 8 + 3 \times 9 =$

   a. 155   b. 45   c. 38   d. 29

23. The cost of 9m of cloth at $42,00 per metre is _____

   a. $51   b. $4,70   c. $368,00   d. $378,00

24. $\frac{6}{9} \times 2\frac{1}{4} =$

   a. $\frac{1}{6}$ b. $\frac{1}{3}$ c. $1\frac{1}{2}$ d. $2\frac{1}{6}$

25. $1111 - 505 =$

   a. 606   b. 909   c. 1616 d. 504

26. What distance does a lorry travelling at 80km/hr cover in 4 hours?

   a. 76km b. 20km   c. 84km d. 320km

27. By how much is 276 greater than 195?

   a. 81   b. 91   c. 121   d. 471

28. $\frac{60}{8}$ as a mixed number in the lowest terms

   a. $6\frac{3}{8}$ b. $6\frac{1}{2}$ c. $7\frac{3}{8}$ d. $7\frac{1}{2}$

29. The change from buying a pair of trousers worth $315 is $85 the money tendered is

   a. $400   b. $230   c. $130 d. $120

30. How long does a lorry take to travel 75km at 100km/h

   a. 2h 55min   b. 1h 20min   c. 45min d. 25min

31. X   What name is given to angle marked X

a. obtuse angle   b. acute angle   c. reflex angle d. right angle

32. 4^3 =   a. 64   b. 43   c. 12   d. 81

33. 3 $^3/_5$kg mealie- meal was put into 4 equal bags. Each bag had.

   a. 800g b. 900g   c. 1000g   d. 1 900g

34.

6cm

The area of the triangle above is 12cm^2 and the base is 6cm. the height of the triangle is _____

   a. 3cm   b. 4cm   c. 6cm d. 2cm

35. The number which is a factor of 162 is _____

   a. 20   b. 14   c. 9   d. 7

36. Mr Gono   travelled 280km in 4 hours. His average speed was

   a. 60km/h   b. 70km/ hr   c. 284km/h d. 1120km/h

37. The diameter of a circle is _____

   a. radius ÷ 2   b. radius x2

   c. circumference x2   d. circumference ÷2

38. In its simplest form $\frac{36}{54}$ is

   a. ¾   b. 4/6   c. 18/27   d. 2/3

39. When a tank is 25% full, it contains 40litres of water. The capacity of the tank is_____

a. 65litres   b. 160litres   c. 150litres

d. 10litres

40.

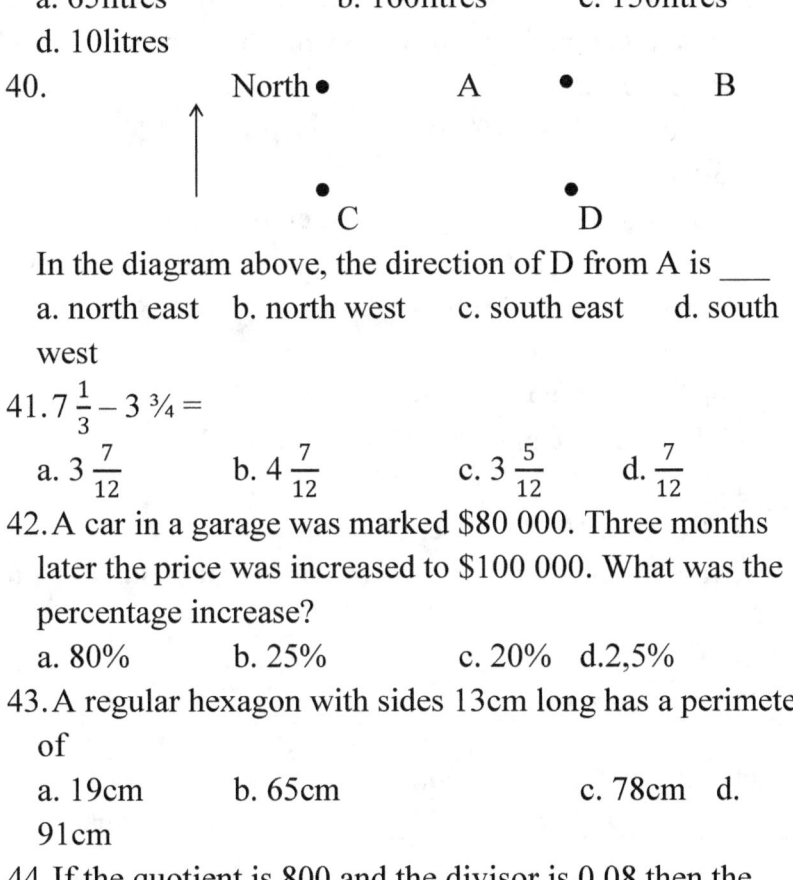

North •   A   •   B

•   •

C   D

In the diagram above, the direction of D from A is ____

a. north east   b. north west   c. south east   d. south west

41. $7\frac{1}{3} - 3\,\tfrac{3}{4} =$

a. $3\frac{7}{12}$ b. $4\frac{7}{12}$ c. $3\frac{5}{12}$ d. $\frac{7}{12}$

42. A car in a garage was marked $80 000. Three months later the price was increased to $100 000. What was the percentage increase?

a. 80%   b. 25%   c. 20%   d.2,5%

43. A regular hexagon with sides 13cm long has a perimeter of

a. 19cm   b. 65cm   c. 78cm   d. 91cm

44. If the quotient is 800 and the divisor is 0,08 then the dividend is____

a. 10000   b. 640   c. 100   d. 64

45. 0,2x0,2x0,2=

a. 0,8   b. 0,6   c. 0,006 d. 0,008

46. Kundai bought a television set for $3000. He later sold it making a profit of 20%. How much did he sell the television set for?

    a. $600   b. $3600   c. $2400 d. $3020

47. Mrs Katema's rectangular garden is 150m long and 120m wide. What is the area of the garden in hectares?

    a. 1,8ha   b. 180ha   c. 18ha d. 1800ha

48. In 2003 Farisai was four years old. Her mother was eight times the age of Farisai, so Farisai's mother was born in _____

    a. 1932 b. 1971   c. 1991 d. 1999

Use the distance table below to answer questions 49 and 50

Juru

721	Jongwe			
94	560	Jiti		
424	600	900	Jana	
125	983	300	184	Jena

49. The longest distance is between

    a. Jongwe and Jena b. Jiti and Jana

    c. Juru and Jiti   d. Jana and Jena

50. A motorist travelled from Jiti to Jena at an average speed of 100km/h. the time taken to travel the journey is

    a. 3h   b. 6h   c. 9h d. 12

SECTION A

1.

   a) State the value of 4 in 364 237   (1)

   b) Change 0, 65 to proper fraction in its lowest terms.

2.

   a) $\frac{1}{4}$ of 36   (2)

   b) $\frac{2}{3} + \frac{1}{6}$ (2)

3. Simplify

   a) 3-0,03=

   b) 0,01x 0,02=

4. 2 ¼ % of $400

5. ¾ - ½ + 1/5=

6. Change 6,4km to metres

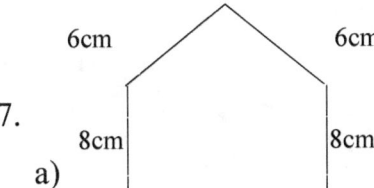

7.

   a)   Name this shape   (1 b)Calculate the perimeter of the shape   (2)

8. Tendai bought a stove for $400 on hire purchase. He paid 25% as deposit and the balance over six months calculate :

   a) Deposit paid   (2)

   b) The simple interest that was charge on the balance at 5% per annum (3)

   SECTION B

9.

   a) A hall has 80 people. ¼ are women. 1/5 are men and the rest are children.

   i. How many women are in the hall?   (1)

   ii. Express the total number of men and women as a fraction of the total people in simplest form.   (1)

10. The graph shows a journey by car from Harare to Karoi which is 420km (not actual). The car had a puncture at 2 pm.

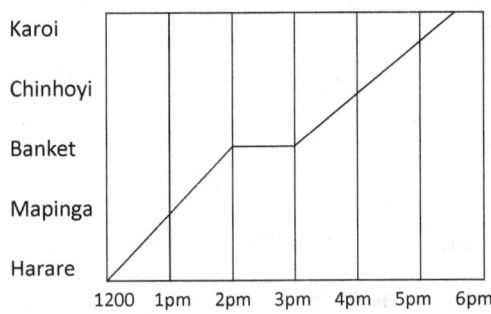

a) State the time taken to mend the puncture (1)

b) Calculate the actual time taken to travel from Harare to Karoi (excluding time for the puncture). (2)

c) Find the average speed of the car excluding the time taken to mend the puncture (2)

11. A rectangular pool has a volume of 80m 3. It is 16m long by 1m deep.

Calculate:

a) The width of the pool 80m3 (16m x 1m) (1)

b) Capacity of the pool in litres . (1)

12. The table below shows the number of cattle that were dipped from Monday to Friday

Day	Mon	Tue	Wed	Thur	Fri
Cattle	200	250	350	430	230

a) How many cattle were dipped on Wednesday?   (1)

a) How many more cattle were dipped on Thursday than Monday?   (1)

b) Find the average number of cattle dipped   (1)

13.

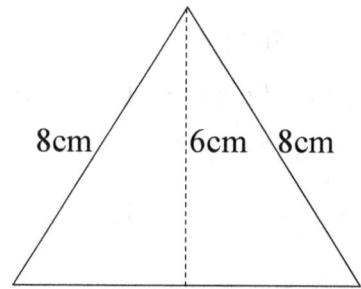

8cm   6cm 8cm

8cm

a) Name the above triangle   (1)

b) Find the perimeter of the triangle.   (2)

c) Find the area of the triangle.   (2)